Illustrated by Veena Mohite

EVERY LAB COAT TELLS A STORY

A compilation of short science fiction stories
about **health, recovery & scientific research.**

DR. PRANITA RAO

INDIA · SINGAPORE · MALAYSIA

Notion Press

Old No. 38, New No. 6
McNichols Road, Chetpet
Chennai - 600 031

First Published by Notion Press 2020
Copyright © Dr. Pranita Rao 2020
All Rights Reserved.

ISBN 978-1-64828-721-3

EVERY LAB COAT TELLS A STORY

A compilation of short science fiction stories about **health, recovery & scientific research.**

DR. PRANITA RAO

INDIA · SINGAPORE · MALAYSIA

Notion Press

Old No. 38, New No. 6
McNichols Road, Chetpet
Chennai - 600 031

First Published by Notion Press 2020
Copyright © Dr. Pranita Rao 2020
All Rights Reserved.

ISBN 978-1-64828-721-3

For Mom, Papa and Priya, with love.

Thank you for everything.

Contents

Preface

Who am I?

I am a part of each one of you who wants to understand the body and the mind. I am someone who wants to acknowledge that science can be fun when learnt using relatable and engaging examples.

What am I doing?

I am attempting to communicate four research topics using four fictitious short stories. Is it going to work?

Well, I will leave you to judge.

Why am I doing this?

During my years as a dental student/dentist, I never felt satisfied with my work. I spent many years feeling frustrated and confused about my career. I liked studying about the human body and mind, but I loved writing about it.

The first time I realized my love for science writing was in my final year of dentistry. I clearly remember sitting in an examination hall, writing long answers for my oral surgery question paper. Answers that I hadn't studied before; answers that just came out of nowhere in the form of words and vivid memories from all the oral surgery lectures. My teachers would appreciate my effort to describe each and every terminology creatively. Looking back at those times, I realize that my subconscious was practising storytelling through those exams. I didn't enjoy practical exams; however, I took great pleasure in weaving stories in my theoretical exams.

Now, I understand why.

Although I completed my dentistry, I understood that I didn't love practising medicine as much as I loved writing about it.

So, here I am. Attempting to live my dream as a science/healthcare writer, hoping, that I can communicate complex medical research topics creatively, as I did in my exam papers.

KINDLY ACKNOWLEDGE – The characters and incidents in this book are purely fictitious and have been developed to resemble realistic events and cases. Any resemblance to any character or scenario would be purely accidental. Texts containing medical facts have been inspired by research papers and the medical terms that appear in a bold format have been further explained in the glossary section.

PLEASE BE AWARE – This book shouldn't be used as a guide to diagnose or treat oneself, instead, to learn about health engagingly. If you do relate your condition to any character, situation or disease mentioned in this book, kindly approach your doctor to discuss it.

That being said, I sincerely hope that you enjoy the book.

Thank you.

Acknowledgements

Like a movie, there are several people responsible for the inception of this book. I might not be able to mention all of those wonderful souls, but if you have been a part of my life (in any capacity), know that in some way, you have motivated me to type a few or many words.

This book is a compilation of short stories about scientific and medical research. This book wouldn't exist without the underappreciated and persevering work of the scientists mentioned at the end of the book. Their work deserves to be known, understood and appreciated by all.

'I believe in you.' The wonders this sentence could do when used in repetition for 26 years. So, I would like to thank my parents for using it all my life.

Thanks to Priya, my little sister, for encouraging me and taking care of me in our childhood. You are my greatest weakness and my biggest strength.

I am immensely grateful to Dr. Raman Khosla and Dr. Bhooshan Shukla for helping me understand and acknowledge the importance of emotional wellbeing. More importantly, I am grateful to you both for being my mentors and well-wishers. I would have barely gotten through life without you both.

I wish to particularly thank Dr. Raman Khosla, who has also been so patient and warm. Thank you for motivating me and helping me identify my passion for writing. You are indeed my blessing. I still remember the first time I'd met you. I was convinced that I was never going to do anything productive with my life. However, you ended up convincing me of the exact opposite. 'You did it, doctor.' You made the impossible, possible. You made this book happen.

A special thanks to all my teachers and friends from Dr. D.Y Patil dental college, Pune for helping me acknowledge the importance of oral health and education.

I don't think I could have written this book, if not for the support and resources that were provided during my master's course by the University of Sheffield, UK. Enrolling in M.Sc Science Communication is by far the best decisions I've made in my life.

Thanks to the staff and students of my master's course for helping me understand science writing, science communication and most importantly, for giving me a place to fit in-something, I couldn't do as a dentist.

I wish to thank all my scicomm mates for inspiring me; more importantly, for coming into my life to direct me towards a different career path. I would particularly like to thank Mae De Los Santos, Benedicte Yende, Tayla Adams and Oriana Trejo for supporting me with my course work and for encouraging me to live my dream.

I would like to thank the staff of Animal and Plant Sciences department.

I would particularly like to extend my gratitude towards my mentor and Science communicator, Millie Mockford for being supportive and for helping me improve my writing and creative ideas.

A special thanks to the NHS trust, Sheffield; Fiona M Boissonade, Emma V Bird and Dr. Simon Atkins for helping me with my dissertation project. I had a foggy vision of being a science writer before working with the Public Patient Involvement (PPI) trust, Sheffield. Working with the PPI trust made me confident and clear about my career goals.

Thanks to Veena Mohite, my illustrator, for being so quick and skilful. It was exciting to work with you; more importantly, I am pleased to have shared the mutual love for art with you.

'If you do not prepare, prepare to fail.' Thank you to my adorable cousins, Akhil Parimi and Rushil Parimi, for using such heavy quotes on me, every time I felt demotivated.

I would like to exclusively thank all the beta readers of this book (Benedicte Yende, Chandni Ahuja, Archana Saptarshi, Rutvik Saptarshi, Mahesh Saptarshi, Maitreyi Saptarshi, Lakmi Peiris and Sunilprasad Shetty) for investing their time and efforts in reading my drafts and giving me their honest reviews.

A big thank you to Notion Press Ltd., India, for helping me publish this book. Your guidance has helped me immensely.

Acknowledgements

A special thanks to Sunilprasad Shetty, who recognized my passion for writing when I couldn't see it in myself. Thanks for planting the 'pen seed' in me. If it weren't for your support, I genuinely do not know if I would have ever had the courage to act on my passion.

Lastly, I would like to thank all my close friends and well-wishers, for being there with me in all my highs and lows. Life has not always been easy for me. Each time I've given up on my dreams, one of you would fly to me and rescue me from the negativity. Thank you for being my angels.

I am not sure how this book is going to do, but writing this book has made me realize the love I have in my life. I hope I can live up to all your expectations.

I LOVE YOU ALL!

(Also, a quick thank you to Cozy Coffeehouse ☕ – An Indie/Folk/Acoustic Playlist | Vol. 1 and Starbucks, Aundh, Pune, for creating the perfect ambience for this budding writer).

Science is often not as complicated as it looks.

Exhibit A-You just have to sit under the right tree at the right time.

1

Tumour Is Not a Rumour

In India, we lose over five people every hour because of oral cancer. Save yourself from oral cancer. Say no to tobacco.

'Not again,' sighed Garima watching Dr. Sidharth Chatterjee, B.D.S, walking towards her with a sullen face. 'Make the call,' said Dr. Chatterjee, nodding his head. Garima, the dental receptionist, took the patient **case paper** from Dr. Chatterjee's hand and quickly turned her attention at the diagnosis section.

'Hey Bhagwaan!' (Oh God! In Hindi)

Garima chanted a quick prayer, as she picked up the phone and dialled the oral and **maxillofacial surgeon**, Dr. Brijesh Iyer.

It was the sixth case of mouth cancer that Dr. Chatterjee had to refer to his senior, just that week.

'Yeh bhi?' (him too?).

Raju kaka, the peon was spooked by Garima's angry facial expressions. He realised that the patient was not too far away from them. 'Bwha-hat hap-pp-end?' inquired Raju kaka in broken English.

'Mouth cancer,' replied Garima in a hushed tone. *'Woh toh dikhra hai, lekin kaunsa cancer?'* (I can see that, but, which cancer?)

'Squamous cell carcinoma,' replied Garima, hastily, trying to ignore the interrogative peon.

'Bhery bad, ah?' (Very bad, ah?)

'Bhery kaka, pura jeebh mein cancer hogaya,' (Very kaka, the cancer has spread to the entire tongue).

'Here valso no, cancer?' (Here also no, cancer?) inquired Rahu kaka, touching his lower lip.

Garima nodded, exhaling deep breaths, imagining the future of the pain-ridden patient from a distance.

'Mr. Bhatt, come.'

Dr. Chatterjee, slowly guided his patient towards the exit door, as he covered his face with a white cloth.

Mr. Bhatt was successful in hiding the 3" cancerous lesion that was protruding through his tongue, almost obliterating the flesh on his cheek. However, the fear in his eyes was detectable by the other patients in the waiting area.

'Will I get better?' asked Mr. Bhatt, wiping his sweat with the white cloth.

'Mr. Bhatt, you must understand you have stage four cancer. However, almost 90% of the oral cancers are squamous cell carcinomas, and Dr. Iyer has treated most of the challenging squamous cell cancer cases in this state. He is the best we have got,' responded Dr. Chatterjee, not making any false promises.

Mr. Bhatt slowly progressed outside *Dantraksha* dental clinic. A tall, young man suddenly struck his attention, he went closer to him and patted his back.

'Put that away, child. If not for this and the 10,000's others that I smoked, I would still be hopeful about living a longer life.'

Mr. Bhatt wore his rubber slippers and slowly, walked away from the young lad, leaving him with an exhausted cigarette butt and some fretful mental images.

2

Social Nirvana

For the one who's heart beats faster than others, when in a crowd.

'I don't think he can handle it.'

Izzie had never seen her husband, Peter looking so pale. While both the middle-aged parents were contemplating about their son's future, Travis, on the other hand, was poking his ear through the gap between his bedroom door.

> *Welcome to Eureka Springs State University. We are pleased to inform you that we have offered you a place in MA Mass communications (Full Time).*

Travis Oracle experienced profuse tremors when he unsealed the envelope containing the offer letter, he received last week. A stir had since then been caused in the Oracle household.

'I will be fine.' Travis's faint voice couldn't be heard by both his parents, who seemed to be writing a long mail to the healthcare centre of the university. 'You don't have to worry about me...or my...condition,' said Travis, raising his voice, blinking copiously.

'Son, we don't doubt your ability, of course,' said Peter, quickly deleting all the tabs on his computer. 'It's just that you might have to be around big groups of people...and... more importantly, the course would require you to... talk,' continued Izzie, with a dreadful expression. As if, talking was Travis's biggest enemy.

'This is the best option for me. Most of the alumni of this course are now celebrities of the journalism field. I will be fine,' said Travis, once again, placing his hand slowly on his father's hand.

He pressed his father's forefinger and reopened the deleted tabs. 'I am not ashamed, dad.'

'Of course not, I just don't want you to worry.'

Travis smiled lightly. His smile was a confirmation to the Oracle's that their son had decided to do something that all three of them considered impossible a few years ago.

* * * *

Within a few weeks, Travis Oracle moved to the student's accommodation at Eureka Springs, Arkansas. It was quite different from New York (a good different, though).

Eureka Springs was quieter, less crowded and greener, which meant that the environment wasn't going to give Travis any health problems. So, he was content, even excited to put back his past behind.

Soon, the classes for mass communication students began at the Rick's University Building.

'Who is up for some icebreakers?' Travis flinched looking at his enthusiastic class moderator, Louise.

'I call this game the speed mates. C'mon, everyone in a circle.' All the 19 students of MA Mass Communications 2018 circled around Louise.

'Now, choose a partner and start introducing yourselves. You can spend 30 seconds with each partner and then, you swap. Comprende?'

Travis froze.

'So many people,' he whispered to himself, feeling like an ant surrounded by many walking humans. He stood still, waiting for someone to come to him. It wasn't that he didn't want to attempt to ask someone, he just, couldn't.

'Hey, are you okay?' Travis thought Louise had heard his pounding heartbeat. He nodded. 'I think you're stuck with me pal.

Looks like you're the odd one out.'

Little did Louise know how accurate that was, considering Travis's situation. However, Louise noticed that Travis didn't seem like the rest of her students. She could tell, as he reminded her of someone.

'So, Travis Oracle, correct?' Travis smiled nervously. 'Just so you know, you wouldn't have been stuck with Louise Campbell if our 20th student had shown up today. I think it's these introduction games that scare them off. If only your course moderator shared the same feelings.'

Louise winked. Travis smiled again, this time, more freely.

The ice had been broken.

* * * *

Although uncomfortably, within a few days, Travis had become a part of the crowd he once dreaded. Often, he would think about the promise he made to his parents. Each time he felt uncomfortable, he reminded himself that it was his conscious choice that led him to his present moment.

'Time for pitches guys. I need you all to give a 3-minute presentation about improving public relations.'

As Louise continued explaining the task, Travis zoned out into the future. He had already foreseen himself, giving a beautifully designed presentation in front of his amazing batchmates while having a frightful **panic attack.**

A panic attack that was time travelling into his present.

'Travis, would you like to say something?' Travis realised that the crowd around him had disappeared, except for Louise. For a few seconds, Travis considered confessing to Louise about his secret, but he hesitated.

Talking about his health with someone new always felt like he was at church, admitting his sins to a priest. Each time he tried opening up, he would witness the respondent making gaunt faces. One even asked him if he was baptized. She thought the demons were attacking him. That one was his best friend who treated Travis like an alien once his secret was out.

'Look, that weirdo is walking towards us. Let's go the other way.'

That was the last time Travis spoke to Bianca Crest, his ex-best friend, who had become his bully by last summer.

So, naturally, Travis became conditioned to adapt only one response, each time he came close to telling someone about his health.

'I am fine,' he said, pacing out of the classroom.

The pacing young lad increased his stride swiftly avoiding eye-contact with anyone who came in his way. He let the floor guide him.

'I need to make an emergency appointment.' 'Welcome to the university health services. How can I help you today?' An automated voice greeted Travis, who was standing outside a big blue door. He repeated himself, only this time, Travis shrieked.

'Hello, I am Erica. Come on in.' The receptionist at the university health services guided Travis to the waiting area. She gave him a pen and a writing pad which held a bunch of papers. Travis quickly grabbed them and glanced at the questions on the documents. He ticked almost every symptom mentioned in them.

Erica looked at Travis with concern and then signalled her colleague at the front desk.

She went closer to Travis and put her hand on the writing pad. 'Umm…you are really lucky.'

Travis didn't understand why, but Erica had caught his attention. 'You see, usually, we do not take walk-ins. However, Dr. Carolina had a cancellation. Would you be comfortable meeting with her?'

'Is she a psychiatrist?' whispered Travis as if he was letting out classified information. 'The best here,' replied Erica.

Travis Oracle soon found himself sitting in a tiny, green room that made him feel all the more uncomfortable. At least now, he was all alone. He felt safer.

'Hi Travis, I am Dr. Carolina Allen. How are you doing?' 'Not very good.' 'Please, do sit,' said Dr. Carolina, watching Travis stand nervously. 'I see you have checked several boxes in the patient form.' Travis remained silent. 'I do have your prior medical history with me, but before we dive into that, would you like to tell me why you are here today?'

'I need to give a presentation next week. I thought I was alright, I even practised some at home, but today, when I was told about it, I…'

'Yes?' Dr. Carolina looked at Travis quivering and thought she might have to take out the tissue box from the drawer. 'All I could think was how I would get humiliated by my teacher and peers next week.'

'What led to that notion, Travis?' 'I could see it.'

'You could see it?' asked Dr. Carolina, hoping that her patient wasn't getting into a **delusional** state. 'I was having racing thoughts about how I would mess up the presentation, and how, I would be negatively evaluated by others. I could even watch my tutor mark me poorly because of my nervous stammer. My chest started to tighten in the middle of the class. Bloody hell! I thought I was going to die of a heart attack.'

'In reality or in your thoughts?'

'Both.' Dr. Carolina put the writing pad aside and folded her arms assertively. 'Travis, I am sorry to hear about what you experienced today. Nevertheless, I am impressed by your response to your panic attack. You are handling yourself well.'

Travis looked puzzled. He did leave his fate of reaching to the university health services to his memory of the tiles on the roads surrounding his campus.

'Certainly! Your first instinct was to get help, which is very different from what happened last summer, Isn't it?'

'You know?'

'Your parents wrote us a pretty detailed summary of your health, which in hindsight, is great! Given how bad your health was last summer, I think that you have come a long way.'

'I have?'

'Of course, now, you have to however understand that your condition isn't something that would vanish in thin air after taking a few medicines. No, it doesn't work that way.'

'I wish it did.'

'Travis, so do I. It's not all that bad though. You are studying at an elite university. You are entitled to some perks. I may not be able to vanish your health condition, but I can certainly decrease your problems surrounding your presentation.'

'How is that?' Dr. Carolina witnessed a glimpse of glee on her patient's face.

'Well, you have given us proof of your health condition, and we understand that your symptoms are genuine. So, I am going to write to your tutor requesting her to not take your presentations skills into consideration, and only mark you for the content of your presentation. You also will be allowed to meet me every week. We can discuss your fears and maybe, start some exercises to improve your health.'

Travis took his eyesight from Dr. Carolina, back to the red-coloured floor. 'Any problem, dear?' 'I thought I was strong enough to not take any help. But once again, here I am.'

'Travis, it's not your fault that you suffer from **social anxiety disorder.** No one talks about the vicious **hormones.'** 'Doctor! I am not a girl,' Dr. Carolina laughed smugly.

'All human beings have hormones, Travis. **Oxytocin,** a hormone found in both sexes has been researched to play a role in causing social anxiety disorders. Also, it can be genetic or hereditary as well. Why would anyone want to purposely feel like you are feeling right now? Now, not accepting how one truly feels is something I find unacceptable. You are giving in to the stigmatism nature of the society, dear. **Mental health** is equally important as physical wellbeing. Would you not go to the doctor if you had diabetes? Would you be ashamed to get help for that?'

It took some convincing and some counselling, but Travis soon agreed to get the help he needed, rather, the support he deserved.

* * * *

Although Travis continued facing challenges in social situations, he managed to not let anything affect his coursework and more importantly, his wellbeing.

'Thomas Paine once said, *"The art of publicity is a black art, but it has come to stay, and every year adds to its potency."* And, that's why we,' paused Travis, 'are important. Thank you.'

Three weeks into the course, he had already begun acing all his presentations and course work.

'Wonderful! Wonderful! I am so happy with this year's work. All of you are doing,' 'wonderful,' screamed all the students.

Louise laughed.

'Anyhow,' said Louise, tilting her wrist, 'she should have been here by now.' Louise got up from her chair and walked towards the door.

'Oh! Why didn't you come in?' The class looked puzzled as they saw Louise poke her head out of the door.

'Guys, meet Biyu Chen.'

'Now, our class is complete,' said Louise.

Travis immediately skipped a heartbeat looking at young Biyu. Was this a new social anxiety symptom?

He wondered.

Louise introduced Biyu to all her classmates, one by one. Biyu Chen kept fidgeting with her lustrous hair that kept falling on her round face. She barely uttered any words. Still, her warm smile and glittery eyes caught the attention of all the men in the class of MA Mass communications.

'Ah! You will like Mr. Oracle here.' 'She will?' squeaked Travis in a high-pitched voice. Louise ignored Travis's nervous outburst.

'Of course, Biyu has done her undergrad thesis in the subject you are interested in as well. *Media platforms and public relation education, correct?*'

Biyu nodded and smiled lightly. Travis folded his hands behind his back like a shy kindergarten boy. Interestingly, so did Biyu.

'Travis, why don't you show Biyu the campus? Our libraries are world-famous,' said Louise grabbing her cycle helmet. 'Ciao now! Until next week, class.'

'This one's going to be one silent love story,' whispered Kirk, Travis's classmate. Travis's heartbeat grew louder and faster as all his mates began to depart.

'I hope you didn't encounter any trouble coming here, from China?' Travis asked fretfully, hoping that she wasn't from another part of Asia. Travis began blinking profusely, watching Biyu browsing something seriously on her phone.

'Is she going to report me for being racist? Is she going to call her boyfriend? Maybe she is trying to show me the shampoo she uses for her shiny hair or maybe, she just wants to avoid me. Does she find me embarrassing? Or just too stupid to be her friend?'

Travis almost felt his heart tear out of his shirt as Biyu lifted her phone towards him. Encounter = Zāoyù, read English to Mandarin translating app on Biyu's phone. Travis exhaled all the air he had collected in his lungs, all at once.

'I did, thank you. Sorry, my English…weak,' said Biyu, feeling embarrassed. For the first time in his life, Travis felt like he had the upper hand in communicating with another human being.

'No problem,' said Travis, immediately downloading the similar app. 'We shall manage,' he said, blushing slightly, showing off his screen to Biyu. Both of them laughed and took off for some sight-seeing.

* * * *

Travis and Biyu walked through the scenic campus of Eureka Spring State university for hours.

What began as a conversation through a translation app, soon turned into a romantic *encounter*. Travis had never felt so alive in his life.

He met someone whom he could talk to without any hesitation, and Biyu, on the other hand, felt the same way about the first foreigner she opened up to. She was taken aback with Travis's patience. Little did she know how much he appreciated people with communication skills that were poorer than his.

'Shall we sit?' asked Biyu, pointing at a quaint corner, near the sports arena. 'Travis,' said Biyu nudging her elbow into Travis's stomach.

'Pink sky!' Travis heard Biyu's high-pitched voice clearly for the first time. Biyu quickly took out her phone and took a photo of it.

'Thank you,' said Biyu, looking into Travis's eyes. Travis didn't blink, at all. 'I feel so comfortable now.'

'You weren't before?' enquired Travis. 'I...I get shy talking to many people at once.' 'Ah! We can be scary. You think it's because of the different people? 'No, no, this happen even in China.'

Travis didn't know what was more charming. Biyu's bad English or the possibility of Biyu having social anxiety.

'I have doctor, in China. I don't know how to tell you...,' said Biyu, tilting her face towards her feet. Travis was quiet. He put his hand in his pocket and took out his phone. Biyu began fidgeting with her hair again, and the colour of her round face began matching the sky colour.

Biyu thought that she might have already embarrassed herself before the one generous person she had met. One she was able to talk to, more importantly, one who was listening to her silence.

'Is he looking for the word stupid in Mandarin. Maybe, he's messaging his girlfriend. Stupid Biyu, you make him uncomfortable. Maybe... I am no good to be his friend. Is it because I am Chinese?' Biyu whispered to herself, feeling despicable.

Travis suddenly broke Biyu's chain of thought and turned his phone screen towards Biyu.

'This is called Jacobson's muscle relaxation therapy. My doctor taught me. It will help.' 'You too? You have Shèjiāo jiāolǜ zhèng?' 'It's called social anxiety disorder here in America,' said Travis quickly switching the phone tabs.

'Aaahhhhh,' gasped Biyu with a broad smile on her face.

She had met her kind.

'Let me show you,' said Travis, equally excited.

'The point of this technique is to let your concentration turn towards your tense muscles. So, you take a hand, close them into tight fists…and, then…release them slowly. Close tightly, open slowly. Now, you do it.'

Biyu began following Travis's instructions. She closed her eyes as Travis instructed. She began taking a few deep breaths and diverted her concentration towards her hands.

She opened and closed her fists and released the remaining anxiety. Travis even demonstrated the same technique using his shoulders and feet.

'Pick your shoulders up and touch them to your ears... hold...now...down! Release!' His demonstration caught the attention of some university students. However, Travis continued.

'Lift your foot slowly, yes good. Now, move your foot up...and...hold...and...down. Can you feel your muscles?'

Biyu only understood half of Travis's instructions, but she did get the gist. She noticed that her mind was getting deviated from anxious thoughts, and her heartbeat was slowing down as she moved her feet up and down. She could feel her breath slowing down as well.

'I think I should sit down,' said Travis, noticing the crowd surrounding them. 'Any better?' Biyu nodded.

'Don't worry, you are not alone. You know 15 million people here in America suffer from social anxiety disorder.'

'You have wise doctor. I was only given these,' said Biyu, showing him her medicine box. 'Hey! I take these too,' said Travis giggling. Biyu didn't respond to him.

'**Zoloft**...a fine social anxiety medicine.' Biyu looked at Travis with widened eyes.

'You doctor too?' 'I mad google user.' Both laughed. 'These pills helped you?' inquired Travis.

'Little bit, yes, but I still feel not so good.' 'I also feel not so good, with only pills...given by my New York doctor. Then I came here, and everything changed.'

'How?' asked Biyu inquisitively. 'You must go to the university mental health services. Dr. Carolina, my doctor here is amazing.'

'Oh, yes?'

'She made me realise the importance of psychotherapy. I was extremely anxious before I met her, then we worked on some solutions and exercises for my triggers. It's called **Cognitive behavioural therapy.** It works wonders when applied with these pills,' said Travis pointing at Biyu's medicine box. However, Travis soon realised that he had used too many complicated words for Biyu to appreciate his mental health knowledge.

'Should we continue the tour?' said Travis, packing his things. 'Okay,' replied Biyu hesitantly.

* * * *

This time, the tour wasn't dreamy. It was just quick, as Travis had only one spot in his mind that he felt was necessary to be shown in his sight-seeing tour.

'Welcome to the university mental health services. How may I be of help today?'

Although Biyu had trouble decoding the foreign woman's accent, she did read the sign on the big blue door.

Biyu wiped her sweaty forehead with her pink scarf and followed Travis beyond the big blue door.

'We would like to make an appointment with Dr. Carolina, please,' requested Travis to Erica, who was gazing at Biyu.

'We?' replied Erica. 'Sorry, she would like an appointment,' said Travis pointing at Biyu, who began sweating even more profusely after hearing her name being called out.

'I think I am okay,' said Biyu quickly, 'don't worry. Dr. Carolina will take care of your social anxiety disorder.'

Travis's elated voice made Biyu's cheeks red. Though Travis was trying to be helpful, he didn't realise that his behaviour was making Biyu feel overwhelmed.

'I am just fine,' responded Biyu, pacing out of the mental health services centre.

Travis looked at Erica feeling embarrassed. 'Did I overstep?' Erica nodded and patted Travis's shoulder. 'You tried to help a friend. It's okay. Not everyone who wants help is ready to get help.'

Travis sighed. He suddenly went back in time and reminisced about the first time he told his parents about his mental health problem.

'There is nothing wrong with him Izzie. This generation cannot handle problems as we did. I lost my dad at the age of five. It left some scars in me, but look at me now, I am all fine.'

Peter was furious with Izzie's suggestion of taking Travis to a psychiatrist. 'Honey, he doesn't sleep on the nights he has social gatherings. Imagine being in a pub with your pals and not being able to respond to their conversations. He gets very conscious, to the extent that he feels that everyone is talking about how hideous he is. Honey, his best friend has turned against him. He's doesn't fall asleep without having at least one massive panic attack each night. It's not healthy. He needs professional help.'

'Oh! C'mon! He just has to man up.'

The walls in the Oracle household were thin. Travis understood that even though he wanted to get help, his father wasn't ready to let him get help.

After fighting for years, Izzie was successful in taking Travis to Dr. Grenndor Park, a famous psychiatrist in New York. Peter never accompanied his son to the appointments until last summer. It wasn't that he didn't love his son, he just couldn't accept the reality.

So, Travis understood Biyu's reaction

'Poor thing,' he whispered as he walked back to his student accommodation. Travis spent the rest of his evening preparing notes for his class.

He took frequent breaks in between to check his phone. *I am sorry*-read the 25th text that Travis wrote and deleted. He didn't know how to approach the situation. Usually, Travis would just apologize to all his 'normal' friends by using his health as an excuse for making them feel uncomfortable. Even though he knew that it wasn't his fault (or his mental health's fault), Travis always had to surrender. He always feared losing more people.

However, Biyu was one of his kind. He didn't know how to respond to someone like him. It was then that he realised how his family felt around him.

'It's like walking on thin ice. I don't know what to do or what not to do. I don't want you to feel even worse.' His cousin brother Jake had once explained Travis, who was inquisitive about knowing how people around him felt when he had a panic attack.

Anxiously, Travis kept all his course work away to deal with his negative thoughts that were making his chest heavier.

He took out the worksheets given by Dr. Carolina that were titled *Be your own cognitive therapist*. She had advised him to use these worksheets whenever he noticed a rise in his anxious thoughts or fears.

He slowly began completing the worksheet while sipping some warm tea.

BE YOUR OWN COGNITIVE THERAPIST

1) Situation (Briefly explain the anxiety-provoking situation)

I loaded Biyu with information about social anxiety (all of that, while flaunting my English) ☹. If that wasn't enough, I forcefully took her to the mental health services. Ugh! She must be crushed.

2) Automatic thoughts

I am so stupid. I acted like a stupid, stupid teenage lover who wanted to show off his skills about his mental health condition. This is a new kind of stupid altogether. I scared her off. I am awful. I broke her heart. She would hate me now. I have lost her...forever.

3) Bodily Sensations

It's that day again when I feel like I might have a stroke. I have had to change my clothes multiple times because I feel uncomfortable and sweaty. There is a big lump in my throat (a feeling of a lump) which I cannot get rid of. I even gargled my throat.

4) Apply 'disputing questions'

Do I know for certain that...? Am I 100% sure that...? What evidence do I have? Can there be another point of view? What is the worst that could happen?

(Adapted from the workbook: Managing social anxiety: a cognitive-behavioural therapy approach; workbook. Oxford: Oxford Univ. Press.)

Continued…

5) Rational Answers

I tried to help Biyu because I didn't want her to go through what I went through. I just wanted her to get easy access to all the help I received from Dr. Carolina.

Yes, I did overdo it slightly, but my intentions and my heart were in the right place. Helping someone isn't stupid. I don't think I did anything awful.

I might have made her feel overwhelmed, but I will apologize for that...and only that. I am not even a 100% sure that she ran off because of something I did. Maybe being in the mental health services centre triggered an anxiety reaction within her. What's the worst that could happen?

She will stop talking to me. Nah! I don't think she will. I don't believe that today was that catastrophic.

I will give her some space, like a good friend ☺.

6) Recheck for bodily sensations

Woah! The lump has disappeared ☺

And...70, 71, 72, 73, 74. Yes! My pulse is back to normal.

(Adapted from the workbook: Managing social anxiety: a cognitive-behavioural therapy approach; workbook. Oxford: Oxford Univ. Press.)

Travis quickly made some copies of his worksheets and shoved them into his bag.

'I will give it to her when she is ready,' he whispered to himself with a light smile.

* * * *

Travis and Biyu became the quietest students in the MA Mass communications classes. Both of them would acknowledge each other's presence by sharing an awkward look each morning. Travis would look at Biyu with a puppy face, and Biyu would reciprocate with a silent frown.

It took Biyu a few days to settle, but she came around one fine day.

'Hi,' said Biyu, patting Travis's shoulder. Travis's day had suddenly turned upside down. 'Hey, your face…it's smiling,' said Travis nervously. Biyu giggled.

'Would you like to have coffee?' Travis nodded multiple times. 'How have you been?' 'Alright, thank you. How about you?' inquired Travis, as both of them departed from their class to the university café.

'I am good, very good, actually.' 'Any particular reason?' Travis's smile disappeared. His mind began flooding with unproductive thoughts.

'I hope there isn't a new guy in the scenario,' whispered Travis. 'What? No,' said Biyu, blushing. 'I met with Dr. Carolina,' 'that's great. How? Why? When?'

'Day before yesterday. I thought about what you said. I…,' paused Biyu, pulling out a chair for herself in the café.

'Really?' 'Yes, everything was happening so soon. You see within one afternoon, I heard so much about my health, and then we went to the mental health place. It was just…' 'Too much?' said Travis, sighing. Biyu nodded. 'I didn't mean to do that. I just wanted to…'

'Help?' 'Yes, you see. It took my family and me several years to get the right doctor. When I found Dr. Carolina, and when I found you, I just didn't want you to wait.'

'That's sweet, but maybe the next time you meet someone like me, you can save a few sight-seeing tours for later?'

Travis smiled.

'Hello. Welcome to Café Rude, what would you like to have?'

'Can I get a decaf americano, please?' asked Travis to the Blonde waitress. 'You mean, may I have a decaf Americano, please?' corrected Biyu, winking.

'My! My! I just skipped a beat,' said Travis, placing his hand on his chest. 'Ahem! Ahem!' 'Sorry, I will have the same as well, thank you,' said Biyu, looking at the irritated waitress.

'You know something, I now have only one mocha latte a day.' 'Why such an atrocity?' enquired Travis, knowing about Biyu's love for her lattes.

'The things Dr. Carolina makes you do,' chuckled Biyu. 'Ah! Yes! Caffeine is an enemy for people like us. I think we have enough adrenaline within us already.'

'Yes, no wonder so many people drink coffee. All to become like us.'

Both laughed.

Months passed by, Biyu and Travis became the best of friends and secret admirers of each other. Each time one of them suffered from a panic attack, the other one would rush to them with a worksheet. Together, they would plan healthy diets, exercise regimes and study sessions.

Biyu even taught Travis a Chinese meditation technique called **Tso-chan-I**, something that Travis and Biyu would practice every day after their classes.

However, as the course was about to end, Travis's anxiety began to pipe once more.

* * * *

'So, this is the last time we might all be together. Ah! I am going to miss you, fools. You all have made me so proud. Here is to all of us. Cheers! Cheers!'

The entire batch of MA Mass communications 2018 raised their glasses along with Louise. The farewell party at the *Rose and Stick* pub was a hoot. Everyone was dancing and singing, except for two.

'Hey, dude! Are you ever going to ask her out?' 'Kirk, I don't think she feels the same way,' said Travis, holding on to his mocktail.

'Are you crazy? Of course, she does. Just ask her out mate. Now is the chance, go for it. She is leaving for China the day after.'

'Ask her out here?' asked Travis, glancing at Biyu, feeling horrified. So many people-he thought to himself. But Kirk was right. If it had to be done, it had to be done there and then. He didn't want to lose Biyu, more importantly, he didn't want to lose her because of his condition. He had let his condition take over many of his aspirations and dreams, but not this one, he couldn't let his condition get in the way this time.

'I have something to say,' said Travis. 'Yo, buddy! Raise your voice. Even I couldn't hear you,' said Kirk, whispering into Travis's ears.

'I have something to say.' A loud squeak interrupted the classes chitter-chatter.

'Travis, what is it?' asked Louise, feeling concerned, hoping he wasn't having a panic attack.

Travis got up and began walking to the opposite end of the table where Biyu was seated like a rock. Her eyes were glued to Travis's. She knew in her heart that he was going to say something. She had been praying for this moment for months.

With every footstep that Travis took, both Biyu's and Travis's heartbeat grew louder and faster. Both felt the need to escape from the room and yet, stay in the room. A part of them wanted to run away as fast as possible, and another part was forcing them to stay in the moment.

'Biyu,' said Travis nervously holding her hand. Biyu's eyes widened as she sighed. 'My life has been difficult. It's not easy, to have a mental health condition and scrape through life while I feel its presence in each sphere of life. My life has been so been so difficult. I've lived each day of my life wishing that nobody has to live a life like mine.

Then, you walked into my life. I still have a mental health condition, and I still have to scrape through life while I feel its presence in each sphere of my life. However, I don't think it matters anymore, because you are there in my life. Now, I wish everyone had a life like mine. There is still darkness, there is still sickness, but there is also…you.'

Biyu slowly wiped her wet face (along with the other girls of her class).

'Social anxiety disorder has kept me away from a lot of things in my life. I am not going to let it keep me away from you. I love you, Biyu Chen.'

Biyu wrapped her arms around Travis. To hell with the crowd-she thought to herself.

The class cheered, including Louise, who had anticipated Travis and Biyu to become close friends (not this close) even before the course had begun.

Biyu had to go back to China for a few months, but she got an internship in New York where she lived happily with her new roommate Travis Oracle. Travis and Biyu continued going to Dr. Grenddor Park, but they stayed in touch with Dr. Carolina as well.

Owing to the therapy sessions, healthy lifestyle, medications, meditation (and each other's company), Travis and Biyu's mental health remained fine.

They would still face their battles, but who doesn't? Their battles were just different from others.

'Bì shā jì.' 'What's that honey?' inquired Travis, while editing a document on his laptop. 'It means nirvana,' 'is that our word for today?'

'That and enormity.'

Travis smiled.

'So, is it ready?' asked Biyu, starring at the laptop. 'Almost.' 'Are you scared?' Travis turned his attention to Biyu and held her hand. 'Will you be there?' 'Of course!' 'Then, no, I am not scared.' 'May, I say you have reached the highest level of social nirvana.'

'Social nirvana? What is that?' laughed Travis. 'A word that I made up. A state wherein one doesn't suffer because of social gatherings.'

'What would I do without you?' Travis chuckled while saving the word file named:

Meet my friend social anxiety-a Ted talk.

3

Boston Brainwashing Effect

'We are such stuff as dreams are made on, and our little life is rounded with a sleep.'

– William Shakespeare

'Two hundred dollars per hour,' uttered a robotic voice. Ahreen pressed multiple buttons on the uncanny intercom device that was stuck to the door. When did A.I. take over this business? Wondered the middle-aged writer hoping for a human to tend her body.

'Yes, let me in. I will pay,' shouted Ahreen pressing her mouth on the intercom.

'Welcome to the Feel-good Restorative Spa. I am Bridget. We don't usually take walk-ins, ma'am. Anyways, how may I serve you today?' Ahreen sighed and forced out the scenes from Chappie from her mind.

Such is an occupational hazard of a writer. They weave more tales in their heads than on paper, but that's the beauty of storytellers. At least, that's what Ahreen Abdella told her parents when she decided to become an author, before dropping out from Boston medical school.

'Madam?' Ahreen felt a cold hand on her shoulder. 'Sorry, I thought I had booked an appointment.'

Flustered, Ahreen scratched her head. 'Anyways, I would like to get a full-body ayurvedic massage, with a female masseuse, someone with good knowledge about the body's **physiology**, please,' replied Ahreen in one breath. 'Of course, this way, please.'

Bridget guided Ahreen through the halls of the spa.

'Pearl is the best we have. In fact, you will be happy to know that she is just a part-time here.' 'Umm…,' 'oh, don't worry,' said the young receptionist looking at the writer's eyes twitch.

'What I meant to say is that she is studying at the university here. She is a **neuropsychology** research student,' whispered Bridget as she led Ahreen into a semi-dark room.

Ahreen felt a sudden aroma of sandalwood gush into her nostrils.

'Here, you can change into these. I hope you feel good,' winked Bridget closing the door after her.

Ahreen quickly got changed and lie flat on the massage table. After a few seconds, the door opened once again.

* * * *

Ahreen was slightly spooked looking at a young, short woman standing in front of her dressed in a blue jumpsuit, looking as weary as herself.

'Welcome, I am Pearl. I will be your masseuse today,' said Pearl, folding her hair behind her ears, hoping to hide the crumbs of fried fish that were spread like a platter on her head.

Ahreen smiled.

(A sign that she could relate to the young masseuse/ researcher).

Pearl dimmed the lights further and put on some light meditative music. 'Could you begin with my back, please?'

'Sure.' 'So, I hear you are a researcher at Boston University.' The weary masseuse glanced at Ahreen, who was lying flat on her stomach and tilting her head upwards. She slowly placed her oily hands on Ahreen's head.

'Please,' whispered Pearl pressing her head slightly towards the massage table.

That was rude, murmured Ahreen. After a few minutes of silence, Ahreen heard the music turn off.

'I have to do something to pay the bills. Hence, a part-time job.' 'Sure, I would wait at the tables in Red's Diner too.' 'You are a medic? From my Uni?'

'Auch!' Suddenly, Ahreen's head and her legs jolted in the air. Her body now looked like a semi-circularly cut watermelon slice.

'Sorry, too much pressure?'

Ahreen nodded, took a few breaths and got back into the conversation. 'I was one. Now, I am a writer.'

'Really?' Why did you quit med school?'

'Money cannot buy happiness; however, my passion can.'

'Wow! Nice! I am happy for you. Tell me something, Miss writer, do all creative people talk so much?' chuckled Pearl, sliding her hands gently over the writer's lower back. '

Well, how else would we get our stories?' Both the client and the masseuse broke out into a burst of soft laughter.

'Anything interesting that you're working on?' 'Sleep!' yelped Pearl. Ahreen raised her eyebrows and looked up.

'Oh, no! I mean, my research team is working on the subject of sleep,' 'Ah!' responded Ahreen pressing her face back on the table. 'Is it another cure to insomnia? Bloody sleep problems!'

'Well, not really. You see, in my opinion, something surreal happens when we are asleep.' 'Hmm, I wouldn't know. Do tell.'

'Okay, imagine you just fell asleep on this table, right here.'

Ahreen gently closed her eyes and let Pearl guide her to another self-brewed story in her mind. 'Now, within a few seconds, blood will flow out, and **cerebrospinal fluid,** the colourless body liquid will flow inside your brain.'

'Yes, I can see it,' replied the writer in a hushed tone. 'Good, now, imagine the cerebrospinal fluid moving in pulses. Let it wash your brain like rhythmically flowing ocean waves.'

Lost in a trance-like state, Ahreen began drawing vivid pictures in her mind. Pearl dipped her hand in warm oil once again and progressed her hand towards Ahreen's neck. With each rub, Ahreen's muscles loosened, and her mind became calmer.

'Now, with every incoming brain fluid wave, the toxins collected in your sleep-deprived brain will move out. Let them wash away all those toxic proteins that impair your memory as well.'

Pearl was indeed an extra-ordinary masseuse. Not only did she manage to soothe Ahreen's tight muscles, but her hypnotizing voice also helped the writer to get a few minutes of peaceful sleep. Something that the writer couldn't do voluntarily.

At the end of 60 minutes of the appointment, Pearl slowly began tapping on Ahreen's head to wake her up.

'My research put you to sleep, ha? My lead scientist wouldn't like to know that.' Pearl smiled at the half-awaken writer and handed her a towel.

After a soothing shower, the writer walked into the waiting area to find the part-time masseuse and Bridget standing with a post-massage mango milkshake glass.

* * * *

'My clients usually look more relaxed. Was something not alright?' enquired Bridget, looking at Ahreen who seemed lost in thoughts. Pearl could feel her heart race.

'Oh! No, the massage was absolutely perfect. I was just wondering how one could detect the cerebrospinal fluid waves. I was trying to recollect the little medical knowledge I had to put the pieces together.' Pearl laughed, and Bridget was now, lost in thoughts.

'We ended up talking science,' said Pearl to Bridget. Pearl guided Ahreen to the sofa in the reception.

'Well, we had help ma'am. You see, thirteen people volunteered to take part in our study. We strapped them with something called the EEG Caps or **electroencephalogram caps**.

The caps look similar to bike helmets, the only difference is that these metal caps carry electrodes. Each person was made to wear the cap and sleep inside a hollow machine called the **Magnetic resonance Imaging** or an MRI machine. This dome-shaped machine uses a magnetic field and radio waves to generate images of the organs and tissues within our body. This way, we could record beautiful images of brain waves.'

'Hmm…,' responded Ahreen gulping the last few drops of the mango milkshake.

She took a few seconds to silently absorb the information she received. 'But why? Why do all this?' 'Well, for one, I just found out today that my study could act as a new sleeping pill,' said Pearl, sniggering.

'More importantly, Ahreen, this study could one day help us discover deeper secrets about various neurological and psychological disorders associated with disrupted sleep patterns like Autism or even Alzheimer's.'

'Interesting! Maybe you could consider me as an option for your next trial,' giggled Ahreen finally handing over the glass. 'Oh, yes, you will even get paid.'

'That's if you get me to sleep.'

'Maybe a massage would help?' replied Pearl.

Ahreen smiled and handed over a two hundred dollar note to Pearl.

The young writer waved goodbye and walked out of the Feel-good Restorative Spa through the E.T operated doors.

Keep my cheque ready. Have an exciting story titled – The feel-good effects of Brainwashing.

Ahreen quickly looked for her editor's name in her phone and pressed the send button.

4

The Blueprint Slicer

Unethical human genetic experimentation violates the moralities of medical ethics. Where there are few disobeying ethical codes, many others, are trying to conserve the sanctity of what remains. CRISPR-Cas9 or any medical technology is like a smoking gun.

The question that remains is whether one uses it to kill the disease within or spread some more.

'Marcus, Marcus, Marcus,' roared the crowd. 'What a fabulous day! Look at those red banners, David.'

George Davis, the sports commentator, jumped out of his seat ecstatically.

'This is quite a sight George, I mean, look at the old lady in blue.' 'The one on the wheelchair you mean? Oh, is that a number eight banner in her hand?'

'Well, it looks like the people of this city already have a hot favourite.' 'You can't blame them, can you? Marcus is incredible! He's the only Ethiopian runner holding the current world record in 5000 meters. He has won many gold medals in the past and not only that,' paused co-commentator David Stanley, trying to adjust his mike. 'He's also gorgeous, isn't he?'

'Well, nobody can ignore that panther-like stride, David.'

'Woah!' bellowed George, 'did Dale just take over Marcus?' 'George, look…the cameras are zooming onto the field now. It looks like Marcus has slowed down considerably.'

'Is he stopping in the middle of the racetrack? He has ten more laps to cover, what on earth is happening?'

'I think he's having trouble breathing George, he looks like he is going to…'

Suddenly, a daunting lull surrounded the Greenfield arena.

In no time, Marcus Bezebah, the country's most loved athletic champion, had hit the ground.

Soon enough, the shocking incidence of the unconscious athlete occupied a place in every journalist's breaking news columns.

'My colleague has just confirmed the news of Marcus having a sudden heart attack. The doctors are refusing to provide us with any information, but...hold on,' paused the presenter from Sports News Daily, adjusting her mike.

'Critical information has come to light in the case of Marcus Bezebah. The doctors are suspecting a condition called **hypertrophic cardiomyopathy**. It is a rare genetic disease that can cause heart attacks during strenuous physical activities.

In my professional opinion, this could be a severe blow to Mr. Bezabah's career and, more importantly, to his life. What lies in the future of this mesmerising human being? The country wants to know, nay, it needs to know. I am Preeti Jagpal, reporting live from St. Theresa's hospital,' concluded the news presenter, cat walking towards the hospital entrance.

St. Theresa's hospital had always been one of the busiest health care centres in the city. However, never had it witnessed such a vast number of masses enter its premises, only to get a glimpse of their brand-new patient.

'Bring me up to speed,' commanded Dr. Sam Mallory, the senior cardiologist. 'Marcus Bezebah, age 32 years, experienced sudden shortness of breath, chest pain, and loss of consciousness,' dictated Intern Gertrude, adjusting her round glasses.

'We did a complete family history examination and a complete **case history** analysis, along with an

echocardiographic assessment. Tests suggest the enlargement of the left ventricle, the heart's largest blood pumping chamber. There are signs of dysfunction of the heart and irregular heartbeats caused by the imbalance between oxygen supply and demand.'

As Gertrude continued narrating her rehearsed test records, Dr. Mallory started pacing towards the reception desk, completely ignoring the presence of his undervalued intern.

'Doctor,' interrupted Brian, one of Dr. Mallory's favourites. 'We did all the tests you ordered,' 'and?'

Dr. Mallory's instant response brought a sudden scowl on Gertrude's face. 'Well, we got the genetics lab involved like you asked. The tests show that he is suffering from hypertrophic cardiomyopathy.'

'Alright, Brian, start with **propranolol**. That drug should help his heart pump blood more slowly, with little force. It should also relieve his pain and discomfort. Advise complete bed rest and set up a meeting with his family. Also,' grunted Dr. Sam Mallory.

'Can someone take care of the damn press; he is a celebrity outside these hospital corridors. In here, he's just another patient. Better keep that in mind,' said the lead cardiologist hinting the overwhelmed junior doctors to keep their calm.

* * * *

'Mrs. Bezebah, the doctor will see you now,' said Dr. Sam Mallory's assistant. Janiah Bezebah seemed to be holding back her tears while walking towards Dr. Sam Mallory's cabin.

'Please take a seat, ma'am. I am sorry about your husband. I hope you are doing alright.'

'Doctor,' whispered Janiah Bezebah, somberly. 'I need to know everything, please, he's all I have.'

'I completely understand. Mrs. Bezebah, as you have been told, Marcus is suffering from hypertrophic cardiomyopathy. It's a genetically inherited disease that can especially, be catastrophic for athletes. Marcus could continue experiencing sudden heart attacks while running. As of now, we are giving him symptomatic medications that will relieve his chest pain and uneasiness. However, I would have to tell you there isn't any permanent cure for his condition.'

'Does this mean he will have to let go of his career?' 'I would recommend that, ma'am.' 'Marcus is going to be devastated. Being an athlete was all he knew.'

Janiah Bezebah couldn't hold back her tears any further. She instantly paced out of Dr. Sam Mallory's clinic, back into the special ward where her husband was lying still. After pouring out for an hour, Janiah sat down on the couch next to Marcus's bed. She took a few deep breaths and rested her head on the couch.

A few minutes later, Janiah heard footsteps approaching the room, but she chose to be non-responsive.

'I can't believe that I am amidst his presence. I was going to buy a bloody expensive ticket just to see his face, thank god for this emergency case. I did save a ton of quid.'

'And he likes this morbid moron,' whispered Gertrude, pulling out the patient chart from one corner of the room.

'Oh! Why didn't you tell me that she would be here?' 'I think she's asleep. You might want to put an end to your callousness if you still want to be Dr. Mallory's puppy,' retorted Gertrude as she sat on a stool opposite the unconscious athlete.

Gertrude always resented Brian, as he was her biggest competitor. His photographic memory and his over-ambitious nature often made him the first one to bag the best cases.

'So, what do you think?' 'About?' 'The obvious question Gertrude, is he going to stop running?'

'Well, he has to, doesn't he? The medication and the surgeries might keep him alive. Still, he will always be at risk for cardiac arrests during strenuous physical activities.'

As both the interns were chatting about the doomed career of Marcus Bezebah, Janiah Bezebah, on the other hand, kept sobbing silently, pretending to be in a deep sleep.

'Anyways,' said Brian jotting down the patient's vitals, 'did you attend Dr. Bridgestone's lecture yesterday afternoon?'

'Ah, the one you missed?' Brian noticed a sudden expression of content on Gertrude's face. 'Yes, I did attend it actually. I found it rather interesting than her usual ones. I was completely attentive in her lecture, for the first time, may I add.'

Brian's eyebrows suddenly rose high up. He was fully aware of his competition. He would often watch Gertrude revise topics a week or two before each lecture.

'Seriously though, I have never seen Dr. Bridgestone teach a lesson so passionately.' 'What was it about?'

'Editing the blueprint of life,' responded Gertrude.

'Sounds bizarre.'

'Just imagine Brian,' answered Gertrude with twinkly eyes. 'Isn't the possibility of having designer babies completely mind-boggling? It's amazing how scientists are using gene-editing technology to knock out unwanted genes.'

'Designer babies? Mankind hasn't reached that far. However, Gertrude, I did hear about the research involved. Isn't it about the gene-editing technology, CRISPR-Cas9?'

'Yes, exactly! Brian, you should have come.' 'Want to bring me up to speed, mate?'

'Well,' sighed Gertrude peeping diffidently at her watch.

'We do have loads of time, Dr. Mallory has told us to monitor Marcus the entire day,' added Brian, hoping to catch on. 'Alright,' groaned Gertrude, secretly rejoicing the opportunity to show off.

'Let's consider the **DNA** sequences in our body as the blueprint to life. Now, according to Dr. Bridgestone, we can tweak the blueprint. All thanks to CRISPR-Cas9!'

'Hmm...sounds like a weapon of some sorts,' grinned Brian, picturing the gene-editing technology in the hands of his favourite superhero.

'It's a little more complicated than getting bit by an insect.'

'Is it really?' Brian's smug smile struck Gertrude's ego like a lightening.

'Oi! Don't be so sour. Okay, tell me, how does this CRISPR-Cas9 work.'

Brian controlled his laughter, watching Gertrude clear her throat; as if, there was an invisible mike in front of her.

'CRISPR is short for Clustered Regularly Interspaced Short **Palindromic Repeats**. These palindromic repeats are nothing but genetic codes in the DNA. Cas9, on the other hand, is like genetic scissors. It is an enzyme that can cut the DNA. Together they form a gene-editing system. The CRISPR part is more like a guiding system that helps the Cas9 locate and severe the defective DNA segments,' narrated Gertrude, doodling on her palms.

'I remember reading something related to bacteria and CRISPR-Cas9. Does that sound right?'

'Aha! That's right, Brian. This genetic tool was found in bacteria as an immune mechanism generated against viral attacks.'

'Kindly, expand,' replied Brian, scratching his head.

'Well, the CRISPR part is the storage house of the memories from past virus attacks. During viral re-infection, the CRISPR part acts as a guide for the genetic scissors, Cas9, to locate and slice the foreign segments.'

'Bacteria's very own defence mechanism, brilliant!'

Brian took a long pause to absorb all the information and quickly started making notes on his phone.

* * * *

On the other end of the room, the intense discussion about the gene-editing technology was making Janiah Bezebah sluggish. She was losing interest and was on the verge of falling off to sleep, for real this time.

'You know,' said Gertrude elbowing her fellow intern. 'We are taught about it because we as doctors could do a lot with this technology.'

'What makes you say that Gertrude?'

'Well, there is a lot of research going on in this hospital, and I hear that they are trying to use CRISPR-Cas9 for treating genetic mutations.'

'Really? What all did Dr. Bridgestone teach in one lecture?'

Gertrude smiled from the corner of her mouth, watching Brain's armhole drench in sweat.

'Well, why did you stop? Tell me the whole thing.'

'It could take all day. I have had two consistent night shifts. I think I will tell the rest later. I need some coffee.'

Gertrude faked a broad, loud yawn and got up from her chair. Brian immediately caught Gertrude's hand and pulled her back.

'Please, tell me everything she told you. This seems complex. I don't even have her notes. What else do you know?'

For the first time, the envious intern felt sympathetic towards her competitor.

'What will I get in return?' bribed Gertrude, feeling uncomfortable with the soft spot that was generating in her heart.

'All my best cases for a month,' replied Brian instantly.

'Deal,' replied Gertrude promptly, happily shaking Brian's hand.

'Where do I begin? Hmm…okay, so, you primarily have to identify the genes that are causing a health problem.

Let me give you an example.' Gertrude took a few seconds to think,' take him, for example,' yelped Gertrude lifting her finger.

'Marcus?' screeched Brian.

Hearing her husband's name, Janiah Bezebah got out of her half-asleep state. For a minute there, she thought that she was dreaming about everything that was happening around her. She opened her left eye wide enough to see the reality, but not wide enough to get caught.

'So, Brian, in Marcus's case, the MYBPC3 is the name of the gene that carries the **mutation** for him to display the disease-hypertrophic cardiomyopathy.

Our brilliant scientists would then study the mutated DNA sequences, following which they would design a specific CRISPR guide.'

'Gertrude, what does this guide actually do?

'According to Dr. Bridgestone, the guide consists of a replica of the mutated genetic sequences. This replica helps in recognizing the target DNA region of interest and directs the Cas9 there for editing.'

'When you say editing…,' paused Brian, who's mind was sketching colourful images of DNA.

'Gene-editing is like the game of Jenga. It's like inserting healthy genes in the DNA and then deleting some harmful or mutated genes.'

'And, thus, changing the entire blueprint.'

'Absolutely!' retorted Gertrude, continuing the intense discussion. 'Now, this complex is introduced to the target cells wherein the target DNA sequences, in this case, the MYBPC3 sequences are located and sliced. At this point, we can knock out the mutated gene and therefore, reverse the incidence of hypertrophic cardiomyopathy.'

'Wow!' sighed Brian with a feeling of instant excitement. Gertrude looked at Brian's amused face and smiled at him. At that moment, both the interns were in a state of mutual awe, overlooking the bitterness between each other.

On the other corner of the room, Janiah Bezebah precipitously recognised the brilliant opportunity to save her husband's life and his career, which was mainly his life. Although she didn't comprehend the medical chitter-chatter, she understood that using CRISPR-Cas9 could save her husband's career and life.

Janiah was just about to break the silence and reach out to the two bright interns who had brought a new ray of hope into her life. Still, she waited for Gertrude to complete her imaginary treatment plan for her husband.

'Isn't that something? I wonder why they haven't used your theory in practice, Gertrude, this man could have continued to run.'

'I hear that there are a lot of researchers trying to use CRISPR-Cas9 for treating diseases like HIV, cancer, **haemophilia, thalassemia** and even hypertrophic cardiomyopathy. However, the success rates aren't great, Brian.'

'Also,' continued Gertrude with an apprehensive look. 'There is always a risk of unwanted gene alterations. These modifications might end up causing more problems than before. Scientists would love to use this in **clinical trials** but getting approval with so many ethical issues would be almost impossible at this stage.'

'Fair enough,' said Brian, pouting at Marcus. 'Alright, I think I just got buzzed into the O.T (operation theatre), I will go now. Perhaps, we could take a coffee break together, Brian. I can call sister Rosie to come in here. I think I overloaded you with a lot of information. That tiny brain of yours cannot take too much, I know.' 'That doesn't explain why I am still a threat to you all,' replied Brian, giggling.

As soon as both the interns left the room, Janiah Bezebah got up swiftly and walked towards her husband. She placed her hand on his head and went close to his ears.

'I found a way my love,' whispered Janiah Bezebah, the woman with an unfathomable mission.

* * * *

'Mr. Sayyed, you most definitely have to exercise. Your cholesterol levels are over the roof. You have to understand, this is serious.'

Janiah Bezebah peeked into Dr. Sam Mallory's cabin. At the same time, the cardiologist was busy instructing his tremendously obese patient about treating exercise as his only medicine. In turn, making Janiah mindful about the irony in the situation.

She barged into Dr. Sam Mallory's cabin without any hesitation. Unlike her previous meeting with Dr. Sam Mallory, this time, Janiah seemed steady. She walked inside the room boldly and pulled out a chair for herself.

'We need to talk, Doctor.' 'I would have appreciated if you could have waited outside until your turn, like all other patients,' said Dr. Sam Mallory in an elevated tone, stressing on the latter part of his sentence.

'Sorry, but I have to talk to you right now, please, I insist.' Instantly, Mr. Sayyed recognised his opportunity and stood up.

'Well, I think I can come back later, take your time madam,' retorted Mr. Sayyed, happily walking out of the heated conversation that he was having with his doctor.

'It's strange how the doctors here have no say,' grunted Dr. Sam Mallory. 'What is it Mrs. Bezebah, how can I assist you?'

'I know that I am not a doctor, but I would like you to consider a different treatment for my husband.' 'Ma'am, was it, Dr. Google, that recommended this "different" treatment plan?'

'No, it's CRISPR-Cas9, the gene-editing technology. I am sure you've heard about it,' said Janiah Bezebah, firmly, hoping that her doctor would finally take her seriously.

Dr. Sam Mallory was the type of person who would very rarely be impressed with people, especially the non-medico population. However, Janiah Bezebah had attained his attention.

'How did you come across it, ma'am,' inquired the rather shocked doctor. 'That's not important. I know I don't have complete knowledge about it.

Nevertheless, I've heard that some doctors are using CRISPR-Cas9 to treat hypertrophic cardiomyopathy. I need you to tell me about all the possibilities.'

'Umm, well,' responded the doctor fidgeting with his stethoscope. 'Mrs. Bezebah, CRISPR-Cas9 is an evolving technology. Yes, it is being used to modify genes and scientists are using this technology to combat diseases at an elementary level. I know for a fact that it has been used to treat deaf mice and infected cattle, but we are talking about Marcus Bezebah here. Mr. Bezebah is not the usual garden rodent, and frankly, I would be heartbroken to see him get worse.'

'You need to understand,' continued Dr. Sam Mallory. 'The effectiveness of CRISPR-Cas9 has not been tested in humans, and there have been many unwanted genetic alterations while using this gene-editing technology.'

'Marcus can end up having more serious problems than he has now, it's dangerous ma'am.' 'I know I am not a doctor, but I am his wife. Marcus may live with the current medications, but the life he would have would kill him from within. Please, is there anything we could do to at least try it under minimal risk conditions?'

'I don't know,' retorted Dr. Sam Mallory nodding his head hesitantly. 'Maybe I can get in touch with Dr. Bridgestone, she would have the knowledge to respond to your proposal. Though, I don't think she would go for it.'

'I understand, but at least it's worth a try if it's in the best interest of my husband.' 'Ma'am,' said Dr. Sam Mallory pointing his pen towards a framed degree certificate. 'Do you know how we get there?'

Janiah was expecting Dr. Sam Mallory to boast about his superior quality of his medical knowledge.

'One would usually think that we are awarded a degree because of the countless number of work hours and the endless number of exams we give. In reality, the worth of the degree is only when we as doctors hold ourselves responsible for another human being's wellness. So, when I tell you that this line of treatment that you're suggesting is dangerous, I am saying that because I know what's in the best interest of your husband.'

Janiah sensed considerable concern in the doctor's face; however, she was convinced about consulting the genetic specialist, Dr. Sabrina Bridgestone.

Janiah left the doctors clinic partly feeling optimistic and somewhat overwhelmed. The senior cardiologist had successfully terrified her regarding the dark side of the gene-editing tool.

She was hoping for someone to validate her actions. Soon enough, her prayers were answered. As she was walking towards the exit, Janiah suddenly witnessed an enigmatic scenario. She saw her guardian angels smiling at her from the reception desk and assumed that she had their blessings to proceed forward. 'Thank you,' yelled Janiah from a distance and promptly, paced out of the hospital wing.

'What was all that about Brian?' 'I don't know, her husband's health might have put her on edge.'

Gertrude looked at her colleague with a despicable look on her face. 'Something's wrong, Brian. She behaved as if we performed some miraculous transformation.' 'Well, strange things happen in this hospital,' sniggered both the interns, not aware of the commotion their little revision session had caused.

* * * *

As Dr. Sam Mallory progressed towards the genetics lab, he started reminiscing about his last meeting with Dr. Sabrina Bridgestone. He vividly remembered each moment he had spent in her presence, as she was the only woman who had instilled the emotions of intimidation within him.

Dr. Sabrina Bridgestone was the most renowned persona in the field of medical genetics. Her track record of conducting outstanding research studies gave her a celebrity status among her colleagues.

'Finally! I have been waiting for you forever.' The senior cardiologist skipped a beat witnessing his dream come true.

'I hope you got the sample I asked for.'

Dr. Sam Mallory stood clueless near the door. He coughed loudly to gain Dr. Bridgestone's attention, who was keenly observing a cell sample under her microscope.

'Goodness! I am utterly sorry, I thought you were my assistant,' sighed Dr. Sabrina Bridgestone. 'Dr. Mallory, isn't it?

'You…you, know me?' retorted the apprehensive cardiologist. 'Oh! of course, please, do come inside.' After exchanging awkward pleasantries, Dr. Sam Mallory gradually explained Marcus Bezebah's case, emphasising on Janiah's obsession with CRISPR-Cas9 technology.

'Dr. Mallory, you do realise that this isn't possible, right? I agree that CRISPR-Cas9 has a lot of potential to combat diseases, but let's face it; we aren't ready for clinical trials. At this point, CRISPR-Cas9 can only be experimented with "diseases in a dish."'

'But doctor,' probed Dr. Mallory. 'I recollected your presentation at the conference last week. You mentioned that CRISPR-Cas9 is being used in an increasing number of experiments, especially involving mammals and invertebrates. You even spoke about the increased efficiency of using CRISPR-Cas9 in treating cancer patients.'

Dr. Sabrina Bridgestone was amazed at the senior cardiologist's memory. Her twinkly eyes and her persistent smile made Dr. Mallory giggle like a small child.

'Yes, but there are still challenges.' 'Says the woman, who wrote a paper on treating Haemophilia. It is a breakthrough for hemophilic patients.'

'Dr. Mallory,' retaliated Dr. Sabrina Bridgestone. 'It is only a breakthrough for hemophilic rats. We aren't ready for human testing. You see, this gene-editing could target healthy DNA sequences that resemble the target mutated DNA sequence. We could end up editing healthy genes, thus, giving rise to unknown and new health problems. It could be catastrophic! It could give rise to additional mutations that could even lead to cell death. Also,' paused the genetic specialist reaching out to her laptop.

'Your timing is just perfect. I was looking at this journal article earlier this morning.'

'Some scientists tried correcting the mutations of MYBPC3 seen in hypertrophic cardiomyopathy. In a fertilized human egg, of course.'

Dr. Sabrina Bridgestone noticed Dr. Mallory's beaming face.

'Don't get your hopes high Dr. Mallory. It says here although $3/4^{th}$ of the mutated cells were altered, but the mosaics were still a huge concern. Mosaics are…,'

'Mosaics are a mixture of normal and mutated sequences that are formed as a result of non-specific gene slicing. I remember that from your presentation,' blushed Dr. Sam Mallory.

'So, I hope that sums up my reasons to why we cannot use CRISPR-Cas9 for treating Mr. Bezebah. Also, we would never get ethical approval, not in time at least. The ethical board would never go for it, and I don't blame them either. This procedure can be perilous.'

A sudden grimace developed on Dr. Sabrina Bridgestone's face.

'We could all end up in prison if this goes south.'

The genetics specialist suddenly became conscious of the senior cardiologist's company and got back to her composed aura.

'We could probably have this conversation again, if and when Bezebah Jr. becomes the matter of concern. He could inherit the disease from his father, you know. Anyways, I hope this was helpful, Dr. Mallory.'

Dr. Sabrina Bridgestone suddenly recognised the presence of a shadow behind her door.

'I am sorry, I need to get back to work,' said Dr. Sabrina Bridgestone, walking back to her microscope in a nervous stride.

'Sure, I am sorry to have interrupted you in the middle of your work.'

'That's alright,' smiled Dr. Bridgestone. She waved her hand hastily, hoping for Dr. Sam Mallory to walk out of the room.

* * * *

Soon after the departure of the senior cardiologist, the shadow behind the door started approaching Dr. Sabrina Bridgestone's table.

'That was close,' whispered the genetics specialist. 'You bet!'

'Phew! Anyways, did you get the sample?' 'Yes, but this feels horrible Dr. Bridgestone.' 'Do I sense a change of heart?'

'Well, I am breaking the law and, so are you.' 'Oh, stop it. This is for science.' 'You have made me take out a cell sample without patient consent. It is unethical!

More importantly, you are going to use your rejected slash modified version of CRISPR-Cas9 technology on this sample. This is unauthorized research. Of course, I will have second thoughts.'

'Listen,' said Dr. Sabrina Bridgestone placing her hand on her partner's shoulder. 'Imagine all the wonders we could do. We could end up making the most effective version of CRISPR-Cas9, and even better, it would be with the world-renowned athlete's sample. We will get famous.'

'You're an insane woman. You think that you have a cure for the mosaics. His wife doesn't even know that you are going to use his mutated cell sample as your puppet to achieve fame. I can't do this, I just…cannot.'

'Listen to me,' roared Dr. Sabrina Bridgestone shutting the door carefully. 'Everything will be forgiven once I get the results. You are already knee-deep in this mess. You better help me get my results now.'

'You're crazy and,' paused the partner, with a sudden expression of elation. 'I think I got whatever I needed,' countered Dr. Sabrina Bridgestone's partner switching off the recorder app in a mobile phone. Dr. Sabrina Bridgestone instantly saw her bright future fade away. She had successfully managed to plan a secretive, error-prone and an illegal procedure.

However, never had she anticipated that an error in judgement could ruin her life.

'You cannot tweak everything with CRISPR-Cas9, can you?' Grunted the furious partner, storming out of the genetics lab. Dr. Bridgestone's companion paced towards the lower level of the hospital to reach the corner-most room.

* * * *

'We need to talk, may I come in?' asked the partner panting heavily. Slowly, the shamefaced partner pulled out the recorder. 'Dr. Mallory, I am extremely sorry,' cried the partner.

'You look mortified. Why don't you sit down, Brian?'

Brian closed Dr. Sam Mallory's cabin not knowing, that his guilty conscious would soon shut the doors of his senior cardiologist's heart.

Glossary

Bacteria	Bacteria are one kind of cell types. Usually a few micrometres in length, bacteria are found in soil, water, organic matter, or plant and animal bodies. Some Bacteria are responsible for causing diseases.
Cardiac arrest	Cardiac arrest is a result of a sudden loss of blood flow when the heart cannot pump blood efficiently.
Case history	A patient's case history is the data collected by a physician by asking specific questions, either from the patient or from other people who know the person and can provide the appropriate information, to obtain useful information in formulating a diagnosis and providing the patient with health care.

Continued…

Cerebrospinal fluid	Cerebrospinal fluid (CSF) is a colourless body fluid located in the spinal cord and brain. This serves as a cushion and provides the brain inside the skull with vital mechanical and immunological defence.
Clinical trials	Clinical trials are clinical research studies. These prospective biomedical or behavioural research studies on human subjects are designed to answer specific questions regarding biomedical or behavioural approaches, including new treatments and proven strategies that require further study.
Cognitive behavioural therapy	Cognitive behavioural therapy (CBT) aims at treating mental health problems. CBT aims to question and change unhelpful ideas, values, and behaviours. It helps develop personal coping strategies which help in solving current issues.

Delusion	A delusion is a firm belief which is not in line with the cultural and educational sense, found on inadequate premises which are not appropriate for rational argument or evidence. This is, as a condition, distinct from a belief based on false or imperfect awareness of thought, dogma, or some other deceptions.
DNA	Deoxyribonucleic acid or DNA is a molecule made of two chains that coil around each other to form a double helix. DNA carries genetic codes responsible for the development, function, growth and reproduction of all known organisms, including many viruses.
Echocardiographic assessment	Echocardiography is an ultrasound test for the heart. This uses two-dimensional and three-dimensional ultra-sonics to create heart images. This test is widely used in diagnosing, treating, and tracking patients with any suspected or confirmed cardiac diseases.
Electroencephalogram	Electroencephalography (EEG) is a brain electrical activity monitoring and recording study. It is typically non-invasive, involving electrodes placed along the scalp.

Continued…

Haemophilia	Haemophilia is a medical condition in which the blood's capacity to clot is severely diminished, causing the patient to bleed excessively, even from the slightest of injury.
Hormones	Hormones are chemical messengers to the body. They travel through the bloodstream to reach various tissues or organs. They work slowly, over time, influencing many different aspects of the body, including growth, metabolism, development, behaviour, mood and chemical reactions.
Hypertrophic cardiomyopathy	Hypertrophic cardiomyopathy is a condition in which a part of the heart is thickened, rendering the heart less able to efficiently pump blood.
Magnetic Resonance Imaging	MRI scanning is a non-invasive and painless technique using a strong magnetic field and radio waves to produce detailed images of the organs and tissues in the body.
Oral and Maxillofacial surgeon	Oral and maxillofacial surgeons are dental surgeons who treat mouth, teeth, jaws, and facial diseases, abnormalities, accidents, and aesthetics.

Mental health	Mental health, is a part of general well-being in which each person understands his or her capabilities, can manage the stresses of life, can function productively and can contribute to their own lives and their society (World health organization).
Mutation	In biological sciences, a mutation is the alteration of the genetic sequence in the DNA of an organism or virus.
Neuropsychology	Neuropsychology is a branch of psychology which is associated with the study of how the brain and the rest of the nervous system influence the cognition and behaviors of an individual.
Oxytocin	Oxytocin is a hormone that exists in both males and females. One of its function is to maintain and regulate human behaviors associated with relationships and bonding.
Palindromic repeats	A palindromic sequence is a genetic sequence in a double-stranded DNA or RNA molecule wherein the genetic codes in a particular direction on one strand matches the codes in the same direction on the complementary strand.

Continued…

Panic attack	A panic attack is most commonly associated with a sudden episode of intense fear or anxiety and physical symptoms based on perceived threat rather than imminent danger.
Physiology	Physiology is a segment of medical sciences that is associated with the study of the normal functioning of living organisms and their parts.
Propranolol	Propranolol is a drug that has been used in treating various mental health conditions including, anxiety disorders and stress reactions.
Social anxiety	A chronic health condition in which social interactions might cause anxiety and depression.
Squamous cell carcinoma (mouth)	Squamous cell carcinoma of the mouth is a condition of environmental factors, the significant of which is tobacco. Aside from cigarette smoking, other carcinogens for oral cancer include alcohol, viruses, radiation and ultraviolet light.
Thalassemia	Thalassemias are blood disorders that are inherited and characterized by decreased production of haemoglobin in the blood.

Tso-Chan-I	Tso-Chan-I is a method of Buddhist meditation. It focuses on the self-exploring journey within. According to this approach, Zen's wisdom appears in many forms, from defusing pressure to growing self-consciousness; knowing and embracing one's characteristics; discovering one's own life meaning.
Zoloft	Zoloft is a medicine that is prescribed to treat depression, obsessive-compulsive disorder (OCD), panic disorder, post-traumatic stress disorder (PTSD), social phobia and other phobias.

Resources

1. Tumour Is Not A Rumor

- Varshitha, A.J. 2015. Prevalence of Oral Cancer in India. Journal of pharmaceutical sciences and research. 7(10), pp. 845–848.

- The Oral Cancer Foundation. (2020). Stages of Cancer. [online] Available at: https://oralcancerfoundation. org/discovery-diagnosis/stages-of-cancer/ [Accessed 27 Jan. 2020].

- MSD Manual Professional Edition. (2020). Oral Squamous Cell Carcinoma - Ear, Nose, and Throat Disorders – MSD Manual Professional Edition. [online] Available at: https://www.merckmanuals. com/professional/ear,-nose,-and-throat-disorders/ tumors-of-the-head-and-neck/oral-squamous-cell-carcinoma [Accessed 27 Jan. 2020].

2. Social Nirvana

- Ziegler, C., Dannlowski, U., Bräuer, D. et al. Oxytocin Receptor Gene Methylation: Converging Multilevel Evidence for a Role in Social Anxiety. Neuropsychopharmacol 40, 1528–1538 (2015) doi:10.1038/npp.2015.2

- Sahel Khakpoor, Omid Saed & Azim Shahsavar | Robert W. Booth (Reviewing editor) (2019) The concept of "Anxiety sensitivity" in social anxiety disorder presentations, symptomatology, and treatment: A theoretical perspective, Cogent Psychology, 6:1, DOI:10.1080/23311908.2019.161 7658

- Mhanational.org. (2020). Social Anxiety Disorder | Mental Health America. [online] Available at: https:// www.mhanational.org/conditions/social-anxiety-disorder [Accessed 7 Feb. 2020].

- Hope, D.A., Heimberg, R.G. and Turk, C.L. (2010). Managing social anxiety: a cognitive-behavioural therapy approach; workbook. Oxford: Oxford Univ. Press.

- Joy, F. and Jose, T. (2014). Effectiveness of Jacobson's progressive muscle relaxation (Jpmr) technique on social anxiety among high school adolescents in a selected school of Udupi district, Karnataka state. [online] Nitte.edu.in. Available at: http://www.nitte. edu.in/journal/March%202014/86-90.pdf [Accessed 3 Jan. 2020].

3. Boston Brainwashing Effects

- ScienceDaily. (2019). Are we "brainwashed" during sleep? Cerebrospinal fluid washes in and out of brain during sleep. [online] Available at: https://www. sciencedaily.com/releases/2019/10/191031174650. htm [Accessed 3 Dec. 2019].

4. The Blueprint Slicer

- Antonis Pantazis, et al. Diagnosis and management of hypertrophic cardiomyopathy. 2015. PubMed. Doi:10.1530

  [Online]. https://www.ncbi.nlm.nih.gov/pmc/articles/PMC46 76455/ (Accessed on November 9, 2018)

- Ali J. Marian, Eugene Braunwald. Hypertrophic Cardiomyopathy: Genetics, Pathogenesis, Clinical Manifestations, Diagnosis, and Therapy. 15/9/2017 PubMed. Doi: 10.1161 Corrado D1, Migliore F, Basso C, Thiene G. Exercise and the risk of sudden cardiac death. 31/09/2006. PubMed. Doi:10.1007

- Feng Zhang, et al. CRISPR-Cas9 for genome editing: progress, implications and challenges. 15/9/2018. Human molecular genetics. 23, Issue R1. pp. R40-R46. Doi:10.1093

- Kira.S.Markorova, et al. Evolution and classification of CRISPR-Cas systems. 9/05/2011.Nature reviews biology. 9, pp. 4657–477

- Liquan Cai a, et al. CRISPR-mediated genome editing and human diseases. 16/07/2016. ScienceDirect. Doi.10.1016

- Hong Ma, et al. Correction of a pathogenic gene mutation in human embryos. 02/08/2017. Nature. 548. pp. 413–419

- Xiang Jin Kang. Addressing challenges in the clinical applications associated with CRISPR/Cas9 technology and ethical questions to prevent its misuse. 06/10/2017. PubMed. 8(11), 791–795. Doi: 10.100724

- Philipe. O Szapary. Physical activity and its effects on lipids. November 2003. SpringerLink. Volume 5, Issue 6. pp. 488–493

- Cocco Am. Dr. Google in the ED: searching for online health information by adult emergency department patient. 15/10/2018. PubMed. 209(8). pp. 342–347

- Xue Gao, et al. Treatment of autosomal dominant hearing loss by in vivo delivery of genome editing agents. 20/12/2017. Nature.553, pp. 217–221

- Junghyun Ryu, et al. Use of gene-editing technology to introduce targeted modification in pigs. 29/01/2018. Journal of Animal Science and Biotechnology. Doi: 10.1186

THE END

Thank you for your time.